Volcanoes and Earthquakes
Different Types

by Glen Phelan

Table of Contents

Develop Language 2

CHAPTER 1 What Causes Volcanoes? 4
                Your Turn: Observe 9
CHAPTER 2 What Causes Earthquakes? 10
                Your Turn: Predict 15
CHAPTER 3 Effects of Earthquakes 16
                Your Turn: Interpret Data 19

Career Explorations 20
Use Language to Explain 21
Science Around You 22
Key Words 23
Index 24

DEVELOP LANGUAGE

Some **earthquakes** are so weak that people don't feel them at all. Others are so strong that they cause great damage.

Discuss the different effects earthquakes can cause with questions like these:

What happened to the hotel?

The hotel _____.

Why are the store items on the floor?

The store items are on the floor because _____.

What did the earthquake do to the road?

The earthquake _____.

What are some other effects of earthquakes?

earthquake – a shaking of Earth's surface caused by the sudden movement of rock

broken road

CHAPTER 1

What Causes Volcanoes?

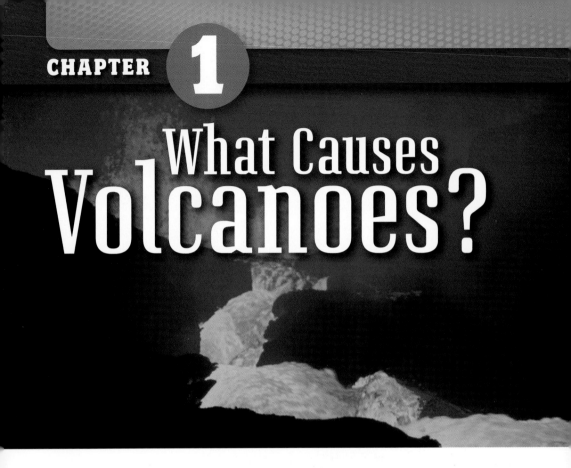

Volcanoes are caused by actions below Earth's surface. A volcano forms when melted rock, called **magma**, collects in an underground magma chamber. When the volcano erupts, the melted rock comes out onto the surface and is called **lava**.

When lava cools, it becomes solid rock. After each eruption, more lava cools. Over time, a mountain of rock can build up. This kind of mountain is called a volcano.

volcanoes – mountains made from melted rock that erupts and hardens

magma – melted rock underground

lava – melted rock that reaches Earth's surface

Sometimes lava flows out gently from a volcano. Sometimes lava explodes from a volcano along with hot gases, chunks of rock, and tiny grains of rock called ash.

▼ Rock layers from many eruptions build up a volcano.

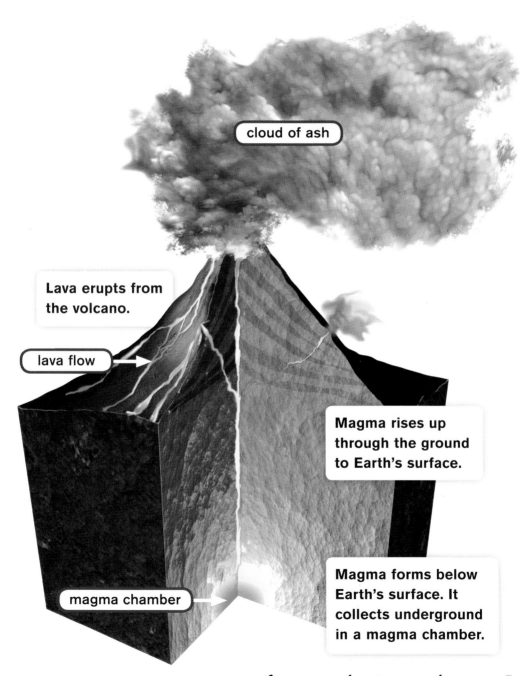

cloud of ash

Lava erupts from the volcano.

lava flow

Magma rises up through the ground to Earth's surface.

magma chamber

Magma forms below Earth's surface. It collects underground in a magma chamber.

Chapter 1: What Causes Volcanoes?

Layers of Earth

Volcanoes are a part of Earth's surface, but magma is deep underground.

Volcanoes form on Earth's **crust**. The crust is the top layer of Earth. It includes all of Earth's land as well as the ocean floors.

The layer beneath the crust is the **mantle**. The top of the mantle is solid rock that floats on a layer of partly melted rock. The rest of the mantle is solid.

The center of Earth is the **core**. The outer core is liquid metal, while the inner core is a ball of solid metal.

crust – the top layer of Earth
mantle – the middle layer of Earth
core – center part of Earth

▼ Earth is made of three main layers.

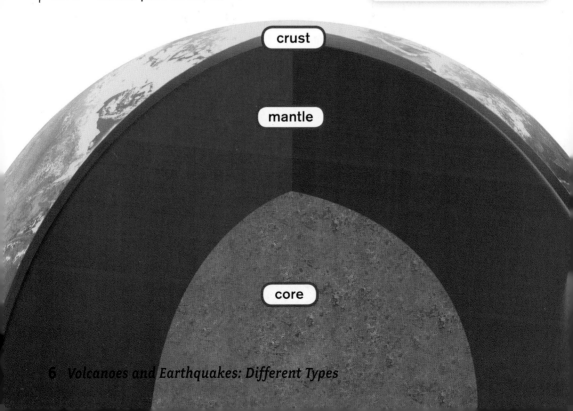

A World of Moving Plates

The crust and top of the mantle form a solid, rocky shell around Earth. But this shell is not one piece. It is broken into about 20 sections called **plates**.

The plates move slowly over the partly melted rock in the mantle. The plates can move into each other, away from each other, or past each other.

If a plate carrying an ocean floor moves into a plate carrying a continent, something interesting happens. The edge of the ocean plate sinks under the continental plate. As it sinks, rocks in the crust melt in the hot mantle. The melted rock can then rise through cracks in the crust and form volcanoes.

plates – large sections of Earth's crust and upper mantle that slowly move

▼ Melted rock rises from a sinking plate and forms volcanoes.

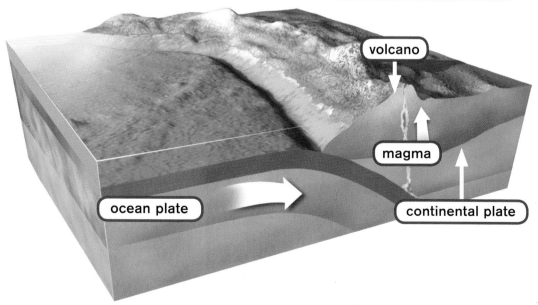

Chapter 1: What Causes Volcanoes?

Three Types of Volcanoes

Volcanic eruptions form three different types of volcanoes. Each type has a different shape. The shape depends on the kind of materials that erupt.

▲ A shield volcano erupts flowing lava. This can form a low, wide mountain.

▲ A composite volcano erupts lava, ash, and large chunks of rock. This can form a tall, cone-shaped mountain.

▲ A cinder cone volcano erupts ash and chunks of rock, but very little lava. This can form a small cone-shaped mountain.

KEY IDEA Volcanoes form when melted rock rises through the crust, erupts onto the surface, and hardens.

YOUR TURN

OBSERVE

Shield volcanoes produce two kinds of lava. Pahoehoe (pah HOH ee hoh ee) lava is hot, thin, and flows quickly. Aa (AH ah) lava is cooler, thicker, and flows more slowly.

Look carefully at the pictures of lava. Then read the groups of words below. Decide which kind of lava each group of words describes.

pahoehoe lava

aa lava

1. rough, jagged, chunky
2. smooth, wavy, wrinkly

MAKE CONNECTIONS

The Pacific Ocean floor is broken into several plates. The edges of these plates that circle the Pacific Ocean are called the Ring of Fire. Use the picture on page 7 to help you explain this name.

USE THE LANGUAGE OF SCIENCE

What is the difference between magma and lava?

Magma is melted rock underground. Lava is melted rock on Earth's surface.

Chapter 1: What Causes Volcanoes?

CHAPTER 2
What Causes Earthquakes?

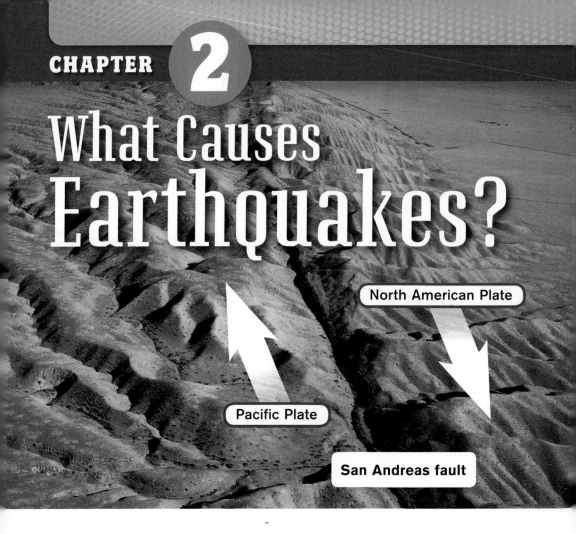

Earthquakes are caused when huge blocks of rock in Earth's crust suddenly move. These blocks move along large cracks called **faults**. The largest faults occur where two plates meet. For example, at the San Andreas fault in California, the Pacific Plate and the North American Plate slide past each other.

The movement of plates creates many smaller faults near the plate boundary. California is crisscrossed by faults.

faults – cracks in Earth's crust along which rock moves

Rock does not move smoothly along a fault. The rough edges get stuck. Motion stops and pressure builds as the blocks of rock continue to push against each other.

When the pressure becomes too great, the blocks of rock break free. They jerk as they suddenly move. The underground location of this sudden movement is the **focus**. The point on Earth's surface directly above the focus is the **epicenter**.

focus – the place in Earth's crust where rocks move suddenly along a fault

epicenter – the place on Earth's surface directly above an earthquake's focus

Explore Language

GREEK WORD ROOTS

epi- (on, over)
+ *kentros* (center)
= epicenter

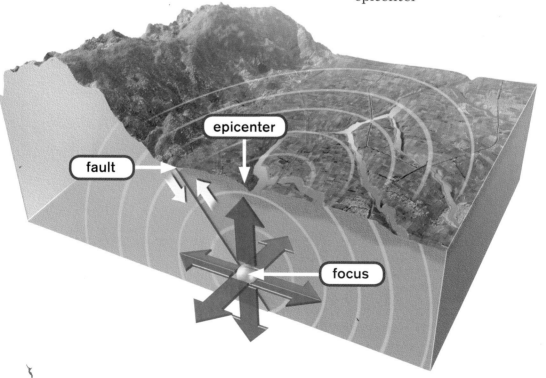

Chapter 2: What Causes Earthquakes?

Seismic Waves

The sudden motion of the rocks at an earthquake's focus creates **seismic waves**. These are waves of energy that travel in all directions from the focus of the earthquake.

An earthquake produces different kinds of seismic waves. Some seismic waves travel through the inside of Earth. Others travel on or near Earth's surface. These are called surface waves. Surface waves cause the greatest damage. They can make the ground roll like an ocean wave.

seismic waves – waves of energy caused by the sudden movement of rocks in Earth's crust

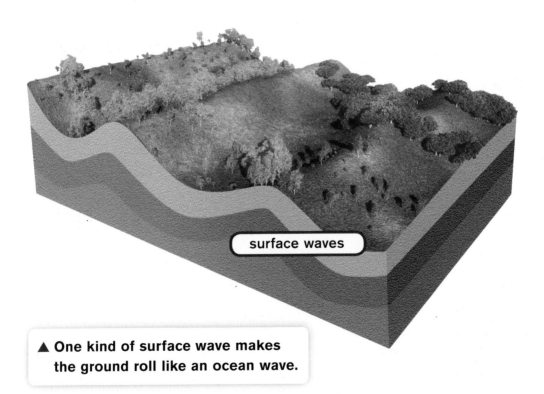

▲ One kind of surface wave makes the ground roll like an ocean wave.

How do scientists observe and study seismic waves? They use **seismographs**. These are instruments that record movements of the ground during an earthquake.

A seismograph has a weight that is attached to a base. During an earthquake, seismic waves shake the base of the seismograph but not the the weight that holds the pen. The pen makes wavy lines that show how the seismic waves are shaking the ground.

Some seismographs record how much the ground moves up and down during an earthquake. Others record how much the ground moves back and forth. This helps scientists figure out the strength and direction of seismic waves.

▲ Scientists study seismic waves to learn about earthquakes and Earth.

seismographs – instruments that record movements of the ground during an earthquake

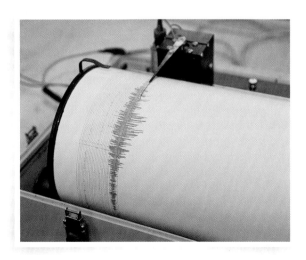

▶ This seismograph records how much the ground moves back and forth during an earthquake.

Chapter 2: What Causes Earthquakes?

Seismographs record tiny movements in the ground from both nearby and faraway earthquakes. Scientists set up seismographs all over the world. With these instruments, scientists can tell when an earthquake happens, where it happens, and how strong it is.

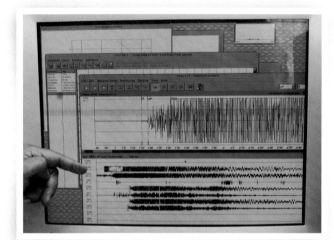

◀ Modern seismographs use computers to measure earthquakes.

By The Way...

One of the first seismographs was invented in China almost two thousand years ago. Eight dragons on a vase each held a metal ball. When the ground shook, a ball dropped into the mouth of a toad, showing the direction the seismic waves came from.

KEY IDEA An earthquake happens when rock suddenly moves along a fault and sends seismic waves in all directions.

YOUR TURN

PREDICT

Look at the map. It shows where faults are located in part of the United States. With a friend, use the map to ask and answer these questions.

1. In what state do you predict the most earthquakes will happen? Explain your prediction.

2. Why are earthquakes unlikely to happen in Minnesota?

MAKE CONNECTIONS

Look again at the map on this page. What do you already know about faults in California?

 STRATEGY FOCUS

Determine Importance

Go over the information about earthquakes. What important ideas are presented? What details support those ideas?

Chapter 2: What Causes Earthquakes?

CHAPTER 3
Effects of Earthquakes

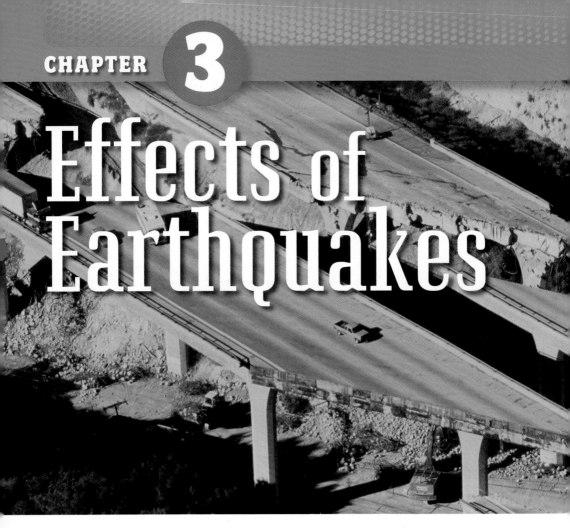

Thousands of earthquakes happen every day. Most are so weak that people cannot feel them. But stronger earthquakes can destroy buildings, bridges, and roads. Usually the greatest damage happens closest to the epicenter.

News reports often describe the strength of an earthquake with a number on the Richter scale. Earthquakes with the largest numbers are the strongest and cause the most damage.

Scientists usually use a more accurate scale to describe the strength of an earthquake. It is called the **moment magnitude scale**. This scale is based on the amount of energy that comes from an earthquake.

On this scale, each higher number stands for 32 times more energy. So an earthquake of magnitude 5 is 32 times stronger than an earthquake of magnitude 4.

Most people cannot feel an earthquake less than magnitude 3. Earthquakes above magnitude 6 can cause great damage. Earthquakes with magnitude 8 or above are rare. They happen only about once a year.

moment magnitude scale – a way of describing the strength of an earthquake based on the amount of energy released

This house was damaged by an earthquake of magnitude 6.

Chapter 3: Effects of Earthquakes

What Are Tsunamis?

Strong earthquakes that occur on the ocean floor can create damaging waves. When an earthquake happens on the ocean floor, part of the floor rises slightly. The rising floor pushes against the water. This push might start a wave called a **tsunami**.

A tsunami travels quickly across the ocean. The wave starts out low—only a few feet high. As it gets close to the coast, however, the wave slows down and the water gets higher. The tsunami might be 30 meters (about 100 feet) tall by the time it hits land!

tsunami – a large wave caused by an earthquake on the ocean floor

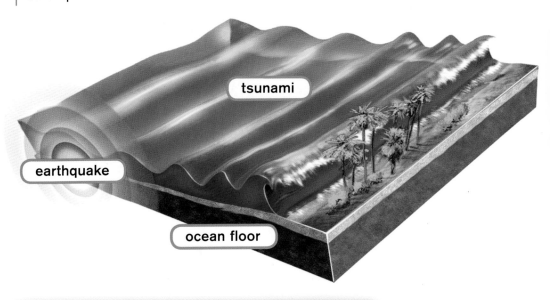

KEY IDEA Most earthquakes are so weak that people cannot feel them. But strong ones can destroy buildings and bridges.

YOUR TURN

INTERPRET DATA

The chart shows the strength of some earthquakes using the moment magnitude scale. Look at the chart with a friend. Ask and answer these questions together.

1. Where did the strongest earthquake happen? How do you know?

2. How many times stronger was the 1923 earthquake in Japan than the 1989 earthquake in Loma Prieta, California? How do you know?

Some Famous Earthquakes

Year	Location	Mag
1906	San Francisco, CA	7.8
1923	Tokyo, Japan	7.9
1960	southern Chile	9.6
1964	Alaska	9.2
1985	Mexico City, Mexico	8.1
1989	Loma Prieta, CA	6.9
1994	Northridge, CA	6.7
1999	Chi Chi, Taiwan	7.6

MAKE CONNECTIONS

When an earthquake happens, why is it important to warn people who live near the ocean?

EXPAND VOCABULARY

Core refers to the central part of something. Find out how **core** is used in each case:

- the **core** of an apple
- the hard **core** of a baseball
- Earth's **core**

Draw pictures of each example. Label each **core** and write a sentence about it.

Chapter 3: Effects of Earthquakes

CAREER EXPLORATIONS

First Responders to the Rescue!

When a major earthquake strikes, people often need help. First responders are the first people to arrive on the scene to help. First responders include fire fighters, emergency medical technicians (EMTs), and police officers.

Fire fighters work to control and put out fires. Earthquakes often break gas pipes. Any spark or flame causes the leaking gas to explode or catch on fire. Fire fighters also use their equipment and skills to rescue people from collapsed buildings.

EMTs provide first-aid and other medical care to people who are injured. They get the injured to a hospital as quickly as possible.

Police officers keep order during the confusion and panic that often happens after an earthquake. Police officers also help rescue people and organize volunteer helpers.

- Would you like to have one of these jobs?
- Tell why or why not.

USE LANGUAGE TO EXPLAIN

Words that Explain

One way to explain is to tell the sequence, or the order, in which something happens. Certain words, such as **first**, **next**, **then**, **eventually**, **after**, and **in the end**, signal sequence.

EXAMPLE

A tsunami wave starts out low. **Then** it slows down and gets higher. **In the end**, a tsunami might be 30 meters tall!

Some phrases show things that are happening at the same time. Some examples are **while**, **as**, **during**, and **at the same time**.

EXAMPLE

As the blocks of rock push against each other, pressure builds.

With a friend, reread the information about volcanoes on pages 4–5. Together, explain what causes a volcano. Use **first**, **while**, and **eventually**.

Write an Explanation

Explain how an earthquake or a tsunami happens.

- Use information from the book and outside sources.
- Use sequence words in your explanation.

Words You Can Use	
first	while
then	after that
by the time	eventually

Science Around You

Earthquake-Proof Bridge Is Now Open

July 16, 2005
Charleston, South Carolina

Over 130 years ago, Charleston, South Carolina had a large earthquake that almost destroyed the city. Now they've built a bridge that is strong enough to handle such a large earthquake.

After four years of construction, the new Arthur Ravenel Bridge is open—ahead of schedule. As part of the opening celebration, thousands of people walked across its two-and-a-half-mile length.

The bridge spans the Cooper River, connecting the cities of Charleston and Mount Pleasant. It is the longest bridge of its kind in North America.

The sleek, modern structure is designed to withstand the forces of nature. It is strong enough to survive the winds of the most powerful hurricanes.

The bridge is also flexible enough to survive the rolling and shaking of an earthquake of magnitude 7.9. This would be similar to the powerful earthquake that nearly destroyed Charleston back in 1886.

Read the newspaper article. Then answer the questions.

- Where is the Arthur Ravenel Bridge located?
- Do you think this bridge will last a long time? Tell why or why not.
- From what you know about earthquakes, do you think there is a fault near Charleston? Explain.

Key Words

core the center part of Earth
Earth's **core** has solid and liquid metal.

crust the top layer of Earth
Earth's **crust** includes all of Earth's land and ocean floors.

earthquake (earthquakes) a shaking of Earth's crust caused by the sudden movement of rock
A strong **earthquake** can change the shape of the land.

epicenter (epicenters) the place on Earth's surface directly above an earthquake's focus
The ground shook the most at the **epicenter** of the earthquake.

fault (faults) a crack in Earth's crust along which rock moves
The movement of plates creates many smaller **faults**.

focus the place in Earth's crust where rocks move suddenly along a fault
Seismic waves travel away from the **focus**.

lava melted rock that reaches Earth's surface
Lava hardens and forms a volcano.

magma melted rock underground
Magma can rise through cracks in Earth's crust.

mantle the middle layer of Earth
The top of Earth's **mantle** is solid rock.

moment magnitude scale a way of describing the strength of an earthquake based on the amount of energy released
On the **moment magnitude scale**, each higher number stands for 32 times more energy.

plate (plates) a large section of Earth's crust and upper part of the mantle that slowly moves
Earth's crust is broken into about 20 **plates**.

seismic wave (seismic waves) waves of energy caused by the sudden movement of rocks in Earth's crust
An earthquake produces different kinds of **seismic waves**.

seismograph (seismographs) an instrument that records movements of the ground during an earthquake
A **seismograph** senses and records seismic waves.

tsunami (tsunamis) a large wave caused by an earthquake on the ocean floor
A **tsunami** might be 30 meters tall by the time it hits land.

volcano (volcanoes) a mountain made from melted rock that erupts and hardens
Every time a **volcano** erupts, more lava collects on Earth's surface.

Index

cinder cone 8

composite volcano 8

core 6

crust 6–7

earthquake 2–4, 10–14, 16–19

epicenter 11

fault 10–11, 15

focus 11

lava 4–5, 8–9

magma 4–5

magma chamber 4–5

mantle 6–7

moment magnitude scale 17, 19

plate 7, 9–10

Richter scale 16

seismic wave 11–14

seismograph 14

shield volcano 8–9

tsunami 18

volcano 4–5, 7–8

MILLMARK EDUCATION CORPORATION
Ericka Markman, President and CEO; Karen Peratt, VP, Editorial Director; Rachel L. Moir, Director, Operations and Production; Mary Ann Mortellaro, Science Editor; Amy Sarver, Series Editor; Betsy Carpenter, Editor; Guadalupe Lopez, Writer; Kris Hanneman and Pictures Unlimited, Photo Research

PROGRAM AUTHORS
Mary Hawley; Program Author, Instructional Design
Kate Boehm Jerome; Program Author, Science

BOOK DESIGN Steve Curtis Design

CONTENT REVIEWER
Tom Nolan, Operations Engineer, NASA Jet Propulsion Laboratory, Pasadena, CA

PROGRAM ADVISORS
Scott K. Baker, PhD, Pacific Institutes for Research, Eugene, OR
Carla C. Johnson, EdD, University of Toledo, Toledo, OH
Donna Ogle, EdD, National-Louis University, Chicago, IL
Betty Ansin Smallwood, PhD, Center for Applied Linguistics, Washington, DC
Gail Thompson, PhD, Claremont Graduate University, Claremont, CA
Emma Violand-Sánchez, EdD, Arlington Public Schools, Arlington, VA (retired)

PHOTO CREDITS Cover and 17 © Visions of America, LLC/Alamy; 1 © Bruno Morandi/Getty Images; 2-3 © AP Images/Wally Santana; 2 © Aerial Archives/Alamy; 3a © AP Images/Huang Ming-Tang; 3b © Stockbyte/Getty Images; 4 © Soames Summerhays/Photo Researchers, Inc.; 5, 6, 7, 11, 12, 18 Illustrations by Chuck Carter; 8a © Krafft/Photo Researchers, Inc.;
8b © Corbis/age fotostock; 8c © Danita Delimont/Alamy; 8d © Photodisc/Punchstock; 9a © Karl K. Switak/Photo Researchers, Inc.; 9b © Photo Resource Hawaii/Alamy; 9c and 9d Lloyd Wolf for Millmark Education; 10 © D. Parker/Photo Researchers, Inc.; 13a © AP Images/David Zalubowski; 13b © Zephyr/Photo Researchers, Inc.; 14a © AP Images/Tatan Syuflana; 14b © SSPL/The Image Works; 15 Map by Mapping Specialists; 16 © David Hume Kennerly/Reportage/Getty Images; 20 © Code Red/Collection Mix/Getty Images; 22 © iofoto/Shutterstock; 24 © Digital Vision/Punchstock.

Copyright © 2008 Millmark Education Corporation

All rights reserved. Reproduction of the whole or any part of the contents without written permission from the publisher is prohibited. Millmark Education and ConceptLinks are registered trademarks of Millmark Education Corporation.

Published by Millmark Education Corporation
7272 Wisconsin Avenue, Suite 300
Bethesda, MD 20814

ISBN-13: 978-1-4334-0086-5
ISBN-10 1-4334-0086-3

Printed in the USA

10 9 8 7 6 5 4 3 2 1

24 *Volcanoes and Earthquakes: Different Types*